SURVEYING MATHEMATICS MADE SIMPLE

An original book by

Jim Crume P.L.S., M.S., CFedS

Co-Authors
Cindy Crume
Bridget Crume
Troy Ray R.L.S.
Mark Sandwick L.S.I.T.

PRINTED EDITION

*****

PUBLISHED BY:
Jim Crume P.L.S., M.S., CFedS

Intersections

Book 8 of this Math-Series

Copyright 2013 © by Jim Crume P.L.S., M.S., CFedS

All Rights Reserved

First publication: November, 2013

Printed by CreateSpace

Available on Kindle and other devices

Cover photo courtesy of public domain. Intersection of a road in Auckland City, New Zealand

TERMS AND CONDITIONS

The content of the pages of this book is for your general information and use only. It is subject to change without notice.

Neither we nor any third parties provide any warranty or guarantee as to the accuracy, timeliness, performance, completeness or suitability of the information and materials found or offered in this book for any particular purpose. You acknowledge that such information and materials may contain inaccuracies or errors and we expressly exclude liability for any such inaccuracies or errors to the fullest extent permitted by law.

Your use of any information or materials in this book is entirely at your own risk, for which we shall not be liable. It shall be your own responsibility to ensure that any products, services or information available in this book meet your specific requirements.

This book may not be further reproduced or circulated in any form, including email. Any reproduction or editing by any means mechanical or electronic without the explicit written permission of Jim Crume is expressly prohibited.

Table of Contents

INTRODUCTION..4

DISTANCE - DISTANCE INTERSECTION..............6

PRACTICAL EXAMPLE 1......................................10

BEARING - DISTANCE INTERSECTION..............13

PRACTICAL EXAMPLE 2......................................16

BEARING - BEARING INTERSECTION................19

PRACTICAL EXAMPLE 3......................................21

SOLUTIONS TO PRACTICAL EXAMPLES............24

CONCLUSION...32

ABOUT THE AUTHOR..33

INTRODUCTION

Straight forward Step-by-Step instructions.

This book is just one part in a series of digital and printed editions on Surveying Mathematics Made Simple. The subject matter in this book will utilize the methods and formulas that are covered in the books that precede it. If you have not read the preceding books, you are encouraged to review a copy before proceeding forward with this book.

For a list of books in this series, please visit:

http://www.cc4w.net/ebooks.html

Prerequisites for this book: A basic knowledge of geometry, algebra and trigonometry is required for the explanations shown in this book.

Book 1 - **Bearings and Azimuths** - How to add bearings and angles, subtract between bearings, convert from degrees-minutes-seconds to decimal degrees, convert from decimal degrees to degrees-minutes-seconds, convert from bearings to azimuths and convert from azimuths to bearings.

Book 2 - **Create Rectangular Coordinates** - How to calculate the northing and easting of an end point given the coordinates of the beginning point utilizing a bearing and distance of a line.

Book 3 - **Inverse Between Rectangular Coordinates** - How to determine the bearing and distance of a line given the coordinates for the beginning and ending point.

Definitions:

Intersections: (a.k.a. Distance-Distance, Bearing-Distance & Bearing-Bearing) The procedure of determining a horizontal position of a point by intersecting lines of directions and/or fixed lengths from two known positions.

DISTANCE - DISTANCE INTERSECTION

Figure 1 shows a typical triangle where all of the distances are known.

For a typical Distance-Distance intersection, the coordinates for points A & B are known and the coordinates for point C is to be determined. The distance between points A & B can be determined by inversing between the known coordinates. The distances for lines AC and BC are also known. The angles A, B and C are to be determined.

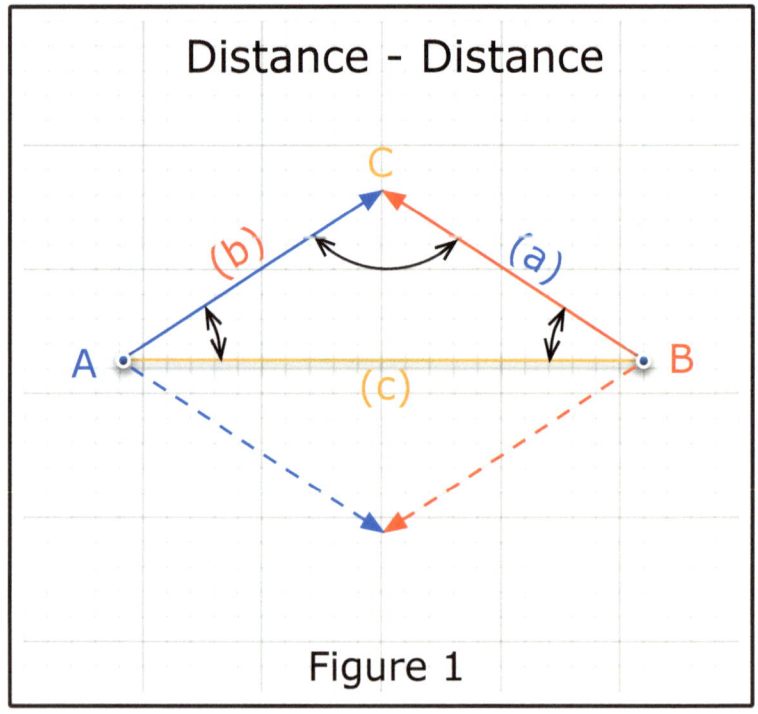

Figure 1

Note: Side 'c' is always the longest side. It is helpful to draw an accurate sketch in order to determine the direction of each of the lines so that you can visually see how to apply the angles to correctly solve the problem.

Two trigonometric laws will need to be utilized to solve the Distance-Distance intersection.

Law of Cosines:

$Cos(A) = (b^2 + c^2 - a^2) / 2 * b * c$

[Solve for A as follows]

$A = ArcCos((b^2 + c^2 - a^2) / 2 * b * c)$

Once you solve for angle A then solve for the remaining two angles using the Law of Sins.

Law of Sins:

$a / Sin(A) = b / Sin(B) = c / Sin(C)$

Solve for angle B:

$Sin(B) = b * Sin(A) / a$

[Solve for angle B as follows]

$B = ArcSin(b * Sin(A) / a)$

Solve for angle C:

$Sin(C) = c * Sin(A) / a$

[Solve for angle C as follows]

$C = ArcSin(c * Sin(A) / a)$

Example 1:

Given:

a = 200.00
b = 175.00
c = 300.00

Solve for angles A, B and C:

A = ArcCos((b² + c² - a²) / 2 * b * c)

A = ArcCos(175.00² + 300.00² - 200.00²) / 2 * 175.00 * 300.00)

A = **39.83815° or 39°50'17.3"**

B = ArcSin(b * Sin(A) / a)

B = ArcSin(175.00 * Sin(39°50'17.3") / 200.00

B = **34.09338° or 34°05'36.2"**

C = ArcSin(c * Sin(A) / a)

C = ArcSin(300.00 * Sin(39°50'17.3") / 200.00

C = **73.93150° or 73°55'53.4"**

Note: If all three angles added together do not equal 180°, then angle 'C' is a supplemental angle and needs to be subtracted from 180° to get the correct angle.

C = 180° - 73°55'53.4" = **106°04'06.6"**

Check: 39°50'17.3" + 34°05'36.2" + 106°04'06.6" = **180°00'00.1"**

Note: Rounding error is dependent upon the number of decimal places that are utilized. It is recommended that at least 5 decimal places be used for all calculations then round the final answer as needed.

All angles must be converted to Decimal Degrees prior to performing trigonometric operations. See Book 1 - "Bearings and Azimuths" for methods on converting Degrees-Minutes-Seconds to Decimal Degrees and vice versa. Also see Book 1 for adding and subtracting bearings and angles.

PRACTICAL EXAMPLE 1

The monument for the corner of Sections 5, 6, 7 and 8 was not found while performing a survey along the North-South line between Sections 5 and 6. Two of the four original Bearing Trees were found. The record distances for these two found Bearing Trees, labeled BT-A and BT-B in **Figure 2** below, will be used to calculate the position of the missing section corner.

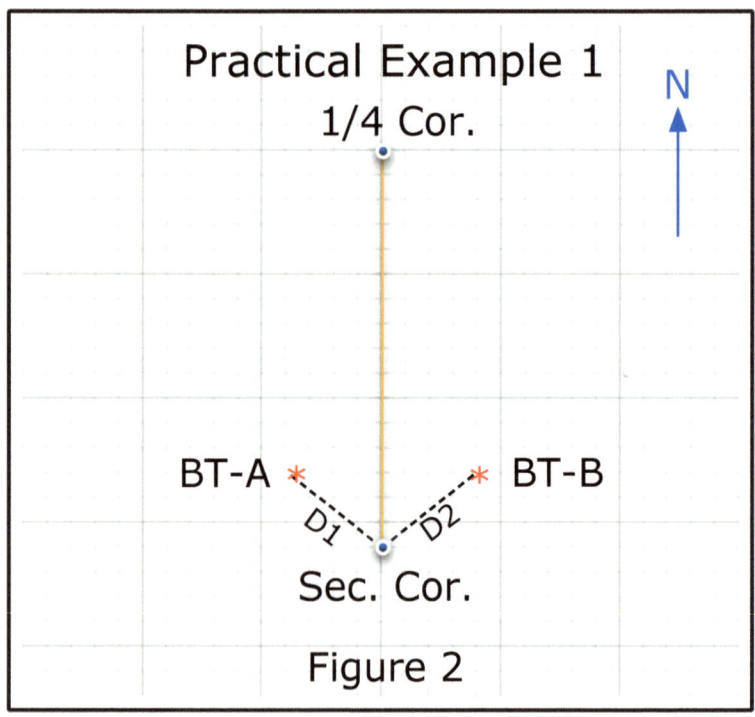

Figure 2

Given:
1/4
N = 1550836.17380

E = 754022.58670

BT-A

N = 1548212.47413

E = 754024.48476

BT-B

N = 1548208.95519

E = 754071.45895

D_1 = 29.1'

D_2 = 35.5'

Solve for the coordinates at the Sec. Cor.:

N (Sec. Cor.) = ???

E (Sec. Cor.) = ???

The solution is towards the end of the book.

NOTES

BEARING - DISTANCE INTERSECTION

Figure 3 shows a triangle where one angle and two distances are known.

For a typical Bearing-Distance intersection, the coordinates for points A & B are known and the coordinates for point C is to be determined. The distance between points A & B can be determined by inversing between the known coordinates. The distance for line AC is known. The angle at B is known. The angles A and C are to be determined. The distance for line BC is to be determined.

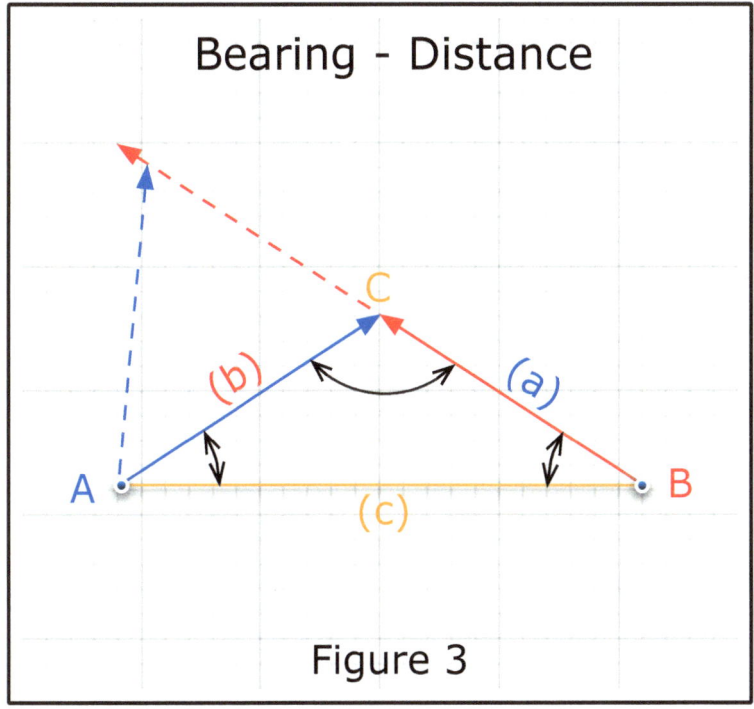

Figure 3

Note: For a Bearing-Distance intersection there are two solutions. It is helpful to draw an accurate sketch in order to determine the direction of each of the lines so that you can visually see how to apply the angles to correctly solve the problem.

The Law of Sins will be used to solve this triangle.

Law of Sins:

a / Sin(A) = b / Sin(B) = c / Sin(C)

Solve for angle C:

Sin(C) = c * Sin(B) / b
[Solve for angle C as follows]
C = ArcSin(c * Sin(B) / b)

Solve for angle A:

A = 180° - B - C

Example 1:

Given:

b = 50.00

c = 100.00

Angle B = 25°00'00"

Solve for angles A, C and distance 'a':

Solution one:

C = ArcSin(c * Sin(B) / b)
C = ArcSin(100.00 * Sin(25°00'00") / 50.00)

C = **57.69729° or 57°41'50.2"**

A = 180° - B - C
A = 180° - 25°00'00" - 57°41'50.2"
A = **97°18'09.8"**

a = Sin(A) * b / Sin(B)
a = Sin(97°18'09.8") * 50.00 / Sin(25°00'00")
a = **117.35040**

Solution two:

For solution two use the supplemental angle for C above.

C = 180° - 57°41'50.2"
C = **122°18'09.8"**

A = 180° - 25°00'00" - 122°18'09.8"
A = **32°41'50.2"**

a = Sin(32°41'50.2") * 50.00 / Sin(25°00'00")
a = **63.91114**

PRACTICAL EXAMPLE 2

You are reviewing a subdivision lot and notice that the distance along the north line is missing. In order to calculate coordinates for the northwest corner of this lot, the missing distance is required. The bearing from the northeast corner to the northwest corner is known.

You have calculated the coordinates for the center of the cul-de-sac and the northeast corner of the lot based upon record information shown on the subdivision plat. You need the missing distance so that you can complete the coordinates for northwest corner of this lot.

Figure 4 shows the subdivision lot configuration. You have calculated the bearing and distance for the line AB. You have the bearing for line BC and the distance for line AC. Calculate the distance for line BC.

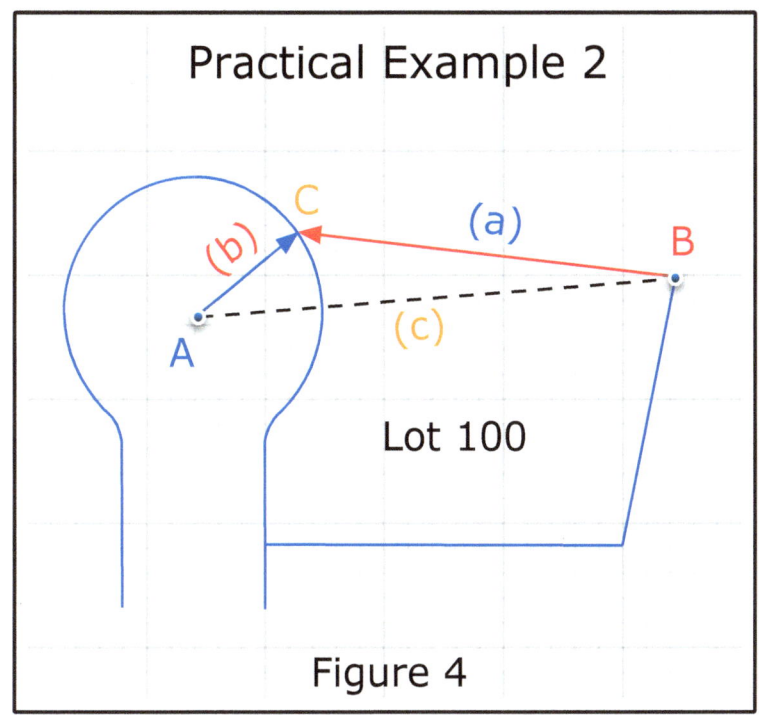

Figure 4

Given:

Line AB = N 86°23'42" E, 198.32'

Bearing of line BC = N 85°10'40" W

Distance of line AC = 50.00'

Solve for the distance for line BC

Distance of line BC = ???

The solution is towards the end of the book.

NOTES

BEARING - BEARING INTERSECTION

Figure 5 shows a typical triangle where all of the angles and one distance are known.

For a typical Bearing-Bearing intersection, the coordinates for points A & B are known and the coordinates for point C is to be determined. The bearing and distance between points A & B can be determined by inversing between the known coordinates. The angles for A, B and C also known. The distance for 'a' and 'b' are to be determined.

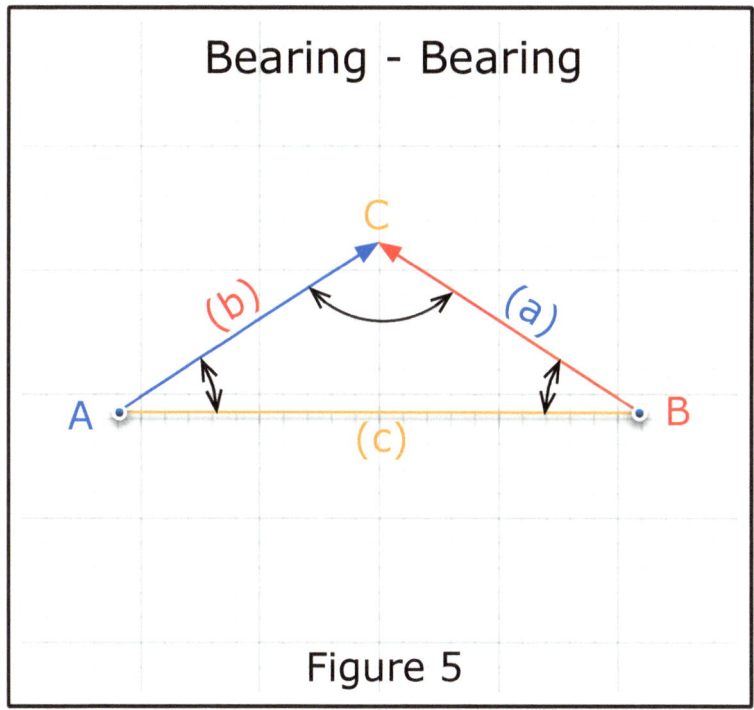

Figure 5

The Law of Sins will be used to solve this triangle.

Law of Sins:

a / Sin(A) = b / Sin(B) = c / Sin(C)

[Solve for distance 'a' as follows]
a = Sin(A) * c / Sin(C)

[Solve for distance 'b' as follows]
b = Sin(B) * c / Sin(C)

Given:
Angle A = 42°28'02.1"
Angle B = 57°21'00.4"
Angle C = 80°10'57.5"
c = 152.95

Solve for distances 'a' and 'b':
a = Sin(A) * c / Sin(C)
a = Sin(42°28'02.1") * 152.95 / Sin(80°10'57.5")
a = **104.80174**

b = Sin(B) * c / Sin(C)
b = Sin(57°21'00.4") * 152.95 / Sin(80°10'57.5")
b = **130.69516**

PRACTICAL EXAMPLE 3

You are reviewing a subdivision lot and notice that the distances along the east and south lines of this lot are missing. In order to calculate coordinates for the southeast corner of this lot, the missing distances are required.

Figure 6 shows the subdivision lot configuration. You have calculated the bearing and distance for the line AB. You have the bearing for lines BC and AC. Calculate the distances for lines BC and AC.

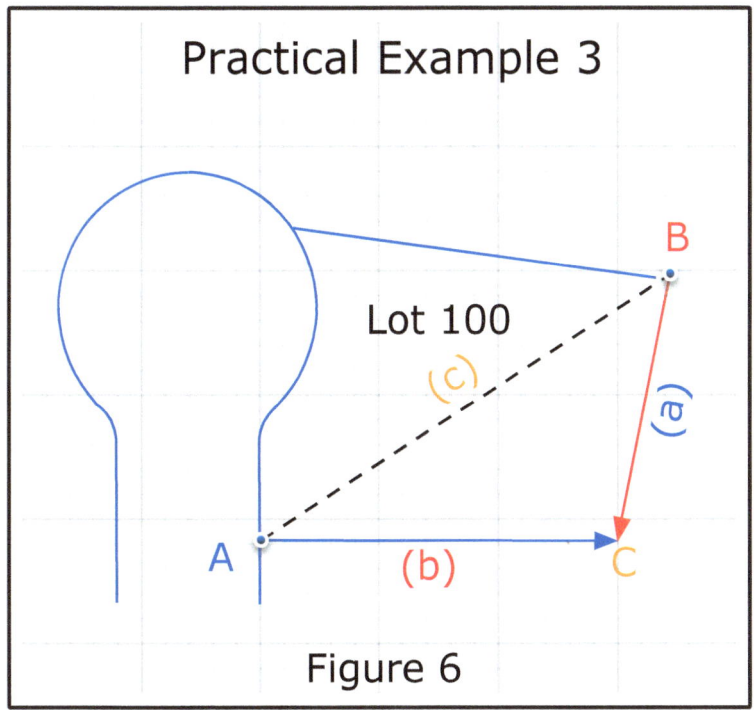

Figure 6

Given:
Line AB = N 49°12'00" E, 125.96'

Bearing of line BC = S 05°11'30" W
Bearing of line AC = N 90°00'00" E

Solve for the distances for lines BC and AC
Distance of line BC = ???
Distance of line AC = ???

The solution is towards the end of the book.

NOTES

SOLUTIONS TO PRACTICAL EXAMPLES

Solution for Practical Example 1:

Given:

1/4
N = 1550836.17380
E = 754022.58670

BT-A
N = 1548212.47413
E = 754024.48476

BT-B
N = 1548208.95519
E = 754071.45895

D_1 = 29.1'
D_2 = 35.5'

Solve for angles at BT-A, BT-B and Sec. Cor.:

Inverse between BT-A and BT-B using the coordinates for these two bearing trees.

Lat = N (BT-B) - N (BT-A)
Lat = 1548208.95519 - 1548212.47413
Lat = **-3.51894**

Dep = E (BT-B) - E (BT-A)
Dep = 754071.45895 - 754024.48476

Dep = **46.97419**

Distance = √(Lat² + Dep²)

Distance = √(-3.51894² + 46.97419²)

Distance = **47.10581**

Bearing = ArcTan(Dep / Lat)

Bearing = ArcTan(46.97419 / -3.51894)

Bearing = **-85.71585° or S 85°42'57.1" E**

Note: See Book 3 - "Inverse Between Rectangular Coordinates" for more details on solving bearings and distances between coordinates..

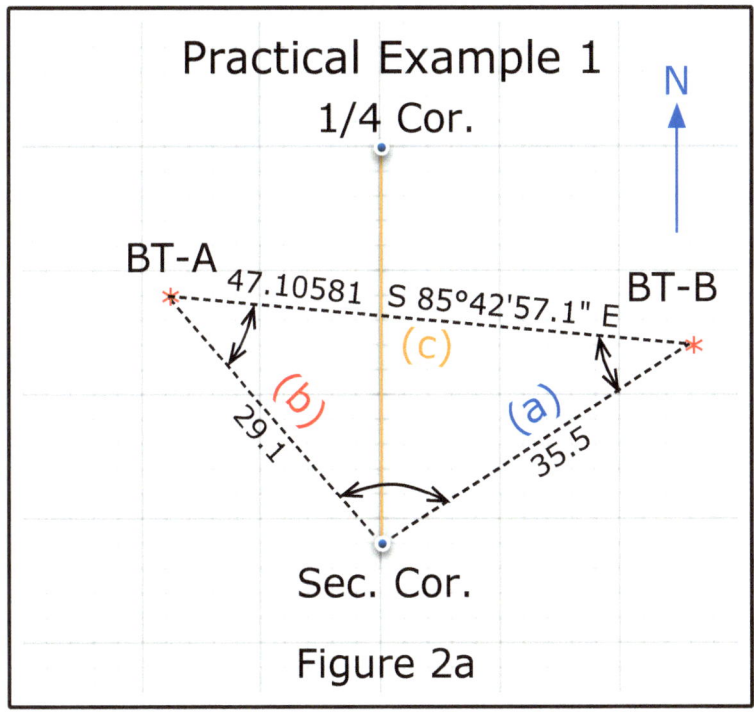

Figure 2a

See **Figure 2a** above for references to the following instructions to solve for the distance-distance intersection for the Section corner.

a = 35.5

b = 29.1

c = 47.10581

BT-A= ArcCos((b^2 + c^2 - a^2) / 2 * b * c)

BT-A = ArcCos((29.1^2 + 47.10581^2 - 35.5^2) / 2 * 29.1 * 47.10581)

BT-A = **48.80884° or 48°48'31.8"**

BT-B = ArcSin(b * Sin(BT-A) / a)

BT-B = ArcSin(29.1 * Sin(48°48'31.8") / 35.5)

BT-B = **38.08658° or 38°05'11.7"**

Sec. Cor. = ArcSin(c * Sin(BT-A) / a)

Sec. Cor. = ArcSin(47.10581 * Sin(48°48'31.8") / 35.5)

Sec. Cor. = **86.89538° or 86°53'43.4"**

Figure 3 indicates that the angle at the Sec. Cor. Should be over 90° therefore the angle calculated above is the supplemental angle. Subtract from 180° to find the correct angle. It is very helpful to draw an accurate sketch in order to determine the direction of each of the lines so that you can visually see how to apply the angles to solve the problem.

Sec. Cor. = 180° - 86°53'43.4" = **93°06'16.6"**

Check : 48°48'31.8" + 38°05'11.7" + 93°06'16.6" = **180°00'00.1"**

Note: Rounding error is dependent upon the number of decimal places that are utilized. It is recommended that at least 5 decimal places be used for all calculations then round the final answer as needed.

Now that we have the angles for the triangle, we can solve for the bearing from either BT-A or BT-B to use for calculating the coordinates for the Section Corner.

We calculated the bearing from BT-A to BT-B to be S 85°42'57.1" E.

We calculated the angle at BT-A to be 48°48'31.8".

Bearing from BT-A to Sec. Cor.

S 85°42'57.1" E - 48°48'31.8" = **S 36°54'25.3" E**

Bearing and distance from BT-A to Sec. Cor.

S 36°54'25.3" E 29.1'

Lat = Cos(Bearing) * Distance
Lat = Cos(S 36°54'25.3" E) * 29.1
Lat = **-23.26868**

Dep = Sin(Bearing) * Distance
Dep = Sin(S 36°54'25.3" E) * 29.1
Dep = **17.47508**

N (Sec. Cor.) = N (BT-A) + Lat

N (Sec. Cor.) = 1548212.47413 + (-23.26868) [SE direction]

N (Sec. Cor.) = **1548189.20545**

E (Sec. Cor.) = E (BT-A) + Dep

E (Sec. Cor.) = 754024.48476 + 17.47508

E (Sec. Cor.) = **754041.95984**

Note: See Book 1 - "Bearings and Azimuths" for instructions on adding and subtracting angles, converting from degrees-minutes-seconds and back. See Book 2 - "Create Rectangular Coordinates" for more details on solving for coordinates given a bearing and distance.

For a check solve the coordinates for the Section corner from BT-B. They should match up with the coordinates above.

The last thing to do is inverse between the Sec. Cor. and the 1/4 Cor. and check that distance with record information.

Solution for Practical Example 2:

Given:

Line AB = N 86°23'42" E, 198.32'

Bearing of line BC = N 85°10'40" W

b = Distance of line AC = 50.00'

Solve for the distance for line BC

Distance of line BC = ???

First solve for the angle at B by subtracting the bearings

B = 180° - N 85°10'40" W - S 86°23'42" W

B = **8°25'38"**

C = ArcSin(c * Sin(B) / b)
C = ArcSin(198.32 * Sin(8°25'38") / 50.00)
C = 35.54118° or 35°32'28.3" [supplemental angle]
C = 180° - 35°32'28.3"
C = **144°27'31.7"**

A = 180° - 8°25'38" - 144°27'31.7"
A = **27°06'50.3"**

a = Sin(A) * b / Sin(B)
a = Sin(27°06'50.3") * 50.00 / Sin(8°25'38")
a = **155.49388**

Solution for Practical Example 3:

Given:

Line AB = N 49°12'00" E, 125.96'
Bearing of line BC = S 05°11'30" W
Bearing of line AC = N 90°00'00" E

Solve for the distances for lines BC and AC

Distance of line BC = ???
Distance of line AC = ???

A = N 90°00'00" E - N 49°12'00" E

A = **40°48'00"**

B = S 49°12'00" W - S 05°11'30" W

B = **44°00'30"**

C = N 90°00'00" E + N 05°11'30" E

C = **95°11'30"**

Check: 40°48'00" + 44°00'30" + 95°11'30" = 180°00'00"

a = Sin(A) * c / Sin(C)

a = Sin(40°48'00") * 125.96 / Sin(95°11'30")

a = **82.64390**

b = Sin(B) * c / Sin(C)

b = Sin(44°00'30") * 125.96 / Sin(95°11'30")

b = **87.87284**

NOTES

CONCLUSION

The above described Distance-Distance, Bearing-Distance and Bearing-Bearing intersections are used quite often on engineering and surveying projects. The above formulas and methods are a valuable resource for your digital library.

ABOUT THE AUTHOR
Jim Crume P.L.S., M.S., CFedS

My land surveying career began several decades ago while attending Albuquerque Technical Vocational Institute in New Mexico and has traversed many states such as Alaska, Arizona, Utah and Wyoming. I am a Professional Land Surveyor in Arizona, Utah and Wyoming. I am an appointed United States Mineral Surveyor and a Bureau of Land Management (BLM) Certified Federal Surveyor. I have many years of computer programming experience related to surveying.

This book is dedicated to the many individuals that have helped shape my career. Especially my wife Cindy. She has been my biggest supporter. She has been my instrument person, accountant, advisor and my best friend. Without her, I would not be the professional I am today. Cindy, thank you very much.

Other titles by this author:

http://www.cc4w.net/ebooks.html

www.ingramcontent.com/pod-product-compliance
Lightning Source LLC
Chambersburg PA
CBHW041147180526
45159CB00002BB/745